Over a Barrel

A Simple Guide to the Oil Shortage

Copyright © 2005 by Tom R. Mast

ISBN: 0-9764440-0-3

Library of Congress Control Number: 2004117812

Submit all requests for reprinting to:
Greenleaf Book Group LLC
4425 Mopac South, Suite 600
Longhorn Bldg., 3rd Floor
Austin, TX 78735
(512) 891-6100

Inquiries should be addressed to:
Tom R. Mast
5714 Painted Valley Drive
Austin, Texas 78759

Published in the United States by
Hayden Publishers
Austin, TX

Cover design and layout by Greenleaf Book Group
Illustrations by Greenleaf Book Group

First Edition

To my wonderful grandchildren—Dylan, Haley,
Hayden, Lauren, and Wyatt

Acknowledgments

I want to express my great appreciation to all those who offered encouragement and detailed advice on the during the creation of this manuscript: Steve Bailey, Ray Bennett, Bob Cox, Pat Hopson, Michael Lo, Bill Mast, Steve Mast, Eddie Mast, Alex McDonald, Chuck Toubin, Mark Williams, and Rick Winter. Special thanks go to Martha Watts and Mary Toubin for their very thorough editing.

Greenleaf Book Group was exemplary in its guidance and work in converting a rough manuscript into a finished product. I especially want to thank Meg La Borde, Allison Pickett, and Hilary Turner.

Last, but not least, I thank my loving wife for her patience during my research and writing.

Contents

Understanding Energy

Understanding Oil

Oil Alternatives

Consequences and Steps for Action

List of Figures

Section 1

Understanding Energy

Chapter 1 | Introduction to the Problem

Oil is currently the most widely used source of energy in the world. Its existence is particularly vital to our transportation industry, which has evolved from the humble horse and buggy to today's sophisticated system in less than 150 years. The facts indicate that world production of oil will start declining in the near future, while demand continues to grow. The ensuing scarcity will inevitably lead to social and economic disruption, unless we develop alternative fuels quickly.

The world must find alternatives for all hydrocarbons, beginning with oil. We have procrastinated until the eleventh hour. Alarmingly, we have not even settled on the technologies for oil alternatives. These new alternatives—and surely there will be more than one—are far from obvious today. Huge technical and social implications are attached to all of the proposed options.

Right now, scientists are unable to pinpoint exactly when the production of oil will fall below the demand, but one thing is for sure: this critical moment will arrive soon, relative to the amount of time that will be required to find replacements for oil.

In his book *Out of Gas*, David Goodstein, distinguished professor at the California Institute of Technology, writes, "Civilization as we know it will come to an end sometime this century unless we can find a way to live without fossil fuels."[1] I am confident that we can find a solution to this problem, but we won't discover an answer unless we work harder and smarter than we have so far.

Alarmingly, we have not even settled on the technologies for oil alternatives.

Changes will come—big changes. What will be our energy source in twenty-five, fifty, or a hundred years? We just don't know, and that is the heart of our dilemma. Kenneth Deffeyes claims in his book *Hubbert's Peak* that in sixty years, production of conventional oil will be one-fifth of the present volume.[2] I can remember sixty years back, and I promise you that it isn't a very long time. I wrote this book hoping to improve the chances that my grandchildren's world will incorporate some viable alternatives for the four-fifths of the oil production that will be gone forever.

There is a reason we find ourselves approaching a time of oil scarcity with no obvious alternatives: a widespread ignorance of energy issues exists, leading to general confusion and apathy. Public apathy discourages policymakers from supporting effective long-range energy programs. Apathy leads people to waste energy in their homes and on transportation.

Will we just accept the devastating effects of the coming oil shortage, or will we actively work to solve the problem while we still may have time? The time to act is now, but it is difficult to be proactive if we don't even understand the basics of energy and oil. Energy's role in our modern world is quite complex, and we can't see the big picture based on the scraps of information gleaned from newspapers, magazines, and television. Reading this book will give you a balanced and factual picture of the medium- to long-range role of oil in supplying the world's energy needs, as well as an understanding of the many technical and social implications of the alternatives to oil. This book is meant as a concise summary of an urgent issue, a synopsis for individuals who want to be conversant with this important topic.

The first objective of this book is to succinctly define energy, oil's role as a source of energy, and the status of alternatives to oil—all succinctly. The second goal is to motivate readers to be more proactive in taking steps to reduce world dependence on hydrocarbon fuels, especially oil. It is certain that our children and grandchildren will enjoy a much less comfortable lifestyle than we do, unless energy issues become a national and international priority soon. Building whole new industries is a mammoth undertaking. We can do it, but the time to begin is today.

Chapter 2 | Energy Concepts

Oil is a source of energy. In fact, it's the number one source in the world.[1] However, before we delve into the specifics of oil, we need to discuss energy and what it does for us.

Energy hasn't changed since my student days, when it was defined in Kent's *Mechanical Engineers' Handbook* as "the capacity for producing an effect."[2] This effect may be heating your home or powering your car down the street or lighting the grocery store. Kent also stated that we should categorize energy as either "stored or in transition."

Stored energy can be classified as kinetic or potential.[3] Kinetic energy is motion; your rolling car has kinetic energy, which is converted into heat when you apply your brakes. Potential energy can take the form of a potential for change in height, chemical bonding, or atomic make-up. Potential energy is the water stored behind a dam. The water has the potential to be converted into mechanical energy and then to electrical energy as it is pulled by gravity through a hydroelectric power generator. Chemical energy is stored in material that will burn, such as oil. Atomic energy is stored in nuclear fuels.

Energy in transition usually takes the form of heat or work.[4] **Heat** is obviously necessary for comfortable, operational homes, schools, and businesses, as well as for activities such as cooking. However, it is also a requirement for most of the work we produce. **Work** is defined as force acting through a distance; a locomotive pulling a train is doing work. Work is exhibited by a propeller pushing a ship or an airplane along, a crane lifting a load, or a steam turbine powering an electrical generator. In these examples, the work results from using a fuel (energy source) to make heat, which is then converted into work by some sort of mechanical engine.

Energy Paths from Source to Use

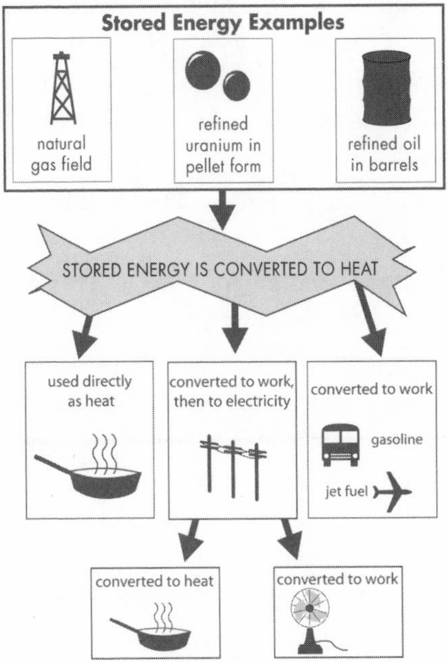

Figure 1.

Energy must have a source. The sun is the originator of all the energy on earth.[5] Most people have heard of solar cells collecting heat directly from the sun's rays to heat water or a home or to generate electrical energy. We are less aware that the sun is responsible for the growth of trees and other biomass that can be used for fuel. The sun's energy also converts organic matter into hydrocarbons—compounds made only of carbon and hydrogen, such as coal, oil, and natural gas—over several hundreds of millions of years. Because we don't have millions of years to wait, we will concentrate on sources that we can convert into useful energy in a much shorter period of time. These sources include coal, oil, natural gas, uranium/nuclear energy, hydroelectric energy, solar power, and wind power. (Electrical energy is a transporter, rather than a source, of energy.)

The source of energy must be moved to the point where it will be converted into heat or work (power). For example, natural gas must be piped to your home to provide heat. It also requires transportation to a central power station for burning in order to make steam, which drives a steam turbine that turns an electrical generator to make electrical power. Oil may be delivered to a refinery by ship, and then delivered to your local station by tank truck after being refined into gasoline. Coal can be moved by rail.

The distribution of power through electrical lines and transformers has some advantages: it's quiet, clean, and doesn't require trucks and drivers. Electrical power is extremely versatile, running many devices, such as your TV, electric drill, computer, dishwasher, lights, air conditioner, alarm system, fans, telephones, hair dryer, radio, doorbell, clock, battery charger, microwave, and thermostat. However, the generation and distribution of electrical power is an inefficient system; much of the power from the source is lost and is no longer available to do useful work. In *Out of Gas*, David Goodstein states that only about one-third of the heat content of a fossil fuel source is turned into electricity, and even more energy is lost in its transmission over long distances.[6] This means that 1 unit of electrical energy used in your home requires about 6 units of energy from the fuel burned at the power plant. It is important that we recognize and understand inefficiencies, because they waste our finite supplies of hydrocarbon sources of energy. In fact, inefficiencies may factor in as serious negatives for some frequently mentioned alternative sources of energy, like hydrogen or ethanol.

Most energy sources arrive at the point of conversion to heat or work in the form of a substance that can be burned. This process chemically combines the matter with oxygen. Examples of these sources include wood and hydrocarbons (fossil fuels). Exceptions are nuclear, hydroelectric, solar, and wind energies.

Energy consumption at the personal level usually involves burning or converting electrical energy into something useful. Cooking, heating with a furnace, and operating your car are examples of consumption via burning. Electrical energy consumption includes lighting your home, powering your electronic devices, and air-conditioning an office. Keep in mind that this electrical energy was generated at a power plant by using a fuel to make heat, which powered a machine turning the electrical generator.

Chapter 3

Energy History

A long time ago, humans didn't even know how to burn wood to cook or to heat their homes. Then they discovered fire. We can only imagine the difference this simple use of energy made to their lifestyle. Later, man learned how to use animals to do the backbreaking work of plowing, transporting, lifting water, and many other tasks. He learned how to use the wind to power ships and how to harness the power of rivers to grind grain.

Hydrocarbon fuels—coal, oil, and natural gas—were formed from the remains of animals and plants on the ocean floor decaying at elevated temperatures under very high pressures for hundreds of millions of years.[1] Coal was the first of these fossil fuels to be used by humans. It was known to the Chinese thousands of years ago. Medieval Europe used some low-grade sulfurous coal for heating. However, for many years, usage was minimal and limited to certain regions.[2]

In 1712, Thomas Newcomen invented a coal-powered machine to pump water up from the English coal mines.[3] This machine had two major effects on the world. First, it permitted excavation in the deep mines, where the higher quality and cleaner burning coal was located. This paved the way for coal to rise above wood as the primary source of heat energy in the developed world. Second, and more important, it demonstrated that we now knew how to make a machine that burned fuel to do useful work.

The Industrial Revolution prospered. James Watt improved the steam engine in 1764, and coal-fired steamships later took the place of sailing vessels.[4] The railroad industry began in the early 1800s and expanded rapidly during the second half of the nineteenth century. Steel mills and other industrial plants multiplied. The invention of the automobile in the latter part of the nineteenth century and of the airplane in the early part of the twentieth century were essential to the industrial arena and our lifestyle today. The entire electrical industry is not much more than a hundred years old, and all of the electronic devices it enables are much younger. Oil shoved aside coal as the world's primary fuel, just as coal had replaced wood.

Useful Inventions Resulting from the Industrial Revolution

Figure 2.

We have burned approximately one-half of all the recoverable oil that was created over a period of 100 million years.

When coal bypassed wood in popularity, and again when oil superseded coal as the world's primary fuel source, neither of the fuels that were being replaced was becoming scarce.[5] The world made each transition because something better came along, not because it had to do so. We are going to be compelled to find alternatives to oil because we have to—on an accelerated schedule—and it is not at all certain at this point that the alternatives will be satisfactory.

In a relatively brief period of time, we have burned approximately one-half of all the recoverable oil that was created over a period of 100 million years.[6] The developed parts of the world with access to sources of energy and the technology that harnesses them have experienced incredible improvements in their standards of living. The population of the world has seen huge growth fueled by energy usage and the lifestyle it enables. However, many researchers think that the world has used its finite supplies of hydrocarbon fuel too aggressively, and that the spurt of growth cannot continue. Some scientists even go so far as to say that Earth's population will decline in parallel with the depletion of hydrocarbon resources and that our standard of living will drop dramatically.

World Energy Demand[7]

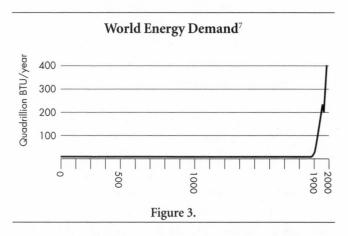

Figure 3.

My mother lived from 1911 until 2002. Her lifetime spanned the development and introduction into mass use of more technology than any other comparable time period in the past. This incredible rate of industrialization made telephones, radios, automobiles, airplanes, airconditioning, televisions, and nuclear power an accepted part of people's lives. She was in awe of the times she had lived through. However, I don't think that my mother fully understood the extent to which all these accoutrements of the good life flowed from our rapid consumption of the energy sources that made them possible.

The current rate at which underdeveloped countries, which contain about 80 percent of the world's population, adopt energy-intensive industrialization is extremely rapid.[8] Instant communication and international commerce and travel are permitting China and India, among others, to go from underdeveloped to industrialized in a mere fraction of the 250–300 years required by Europe and the United States. A quick scan of the newspapers is all it takes to comprehend the increasing pressure this puts on sources of energy, especially oil.[9]

There are many people in the world, particularly in Africa and Asia, who still live in undeveloped countries.[10] At some point, they too will have the education and the resources to demand a share of the energy pie for their economic development, further increasing the demand for energy.

We have harnessed energy to perform an almost unbelievable list of tasks for us over a very short period in human history. In doing so, we have used our convenient hydrocarbon fuels, especially oil, so rapidly that we have not provided for adequate alternatives to follow as our oil supply is depleted.

Chapter 4 | Uses of Energy

We consume energy in four broad categories: residential, commercial, industrial, and transportation.[1] Even though we may relate more easily to the residential and transportation categories, we consume energy from the commercial and industrial categories, too. For example, our lifestyle is improved by the postman's vehicle, the plastic containers in our kitchen, the fresh milk available in our grocery store, and the abundance of manufactured goods in our local hardware store. All of these things are dependent on abundant supplies of energy to government, commercial, industrial, and transportation activities.

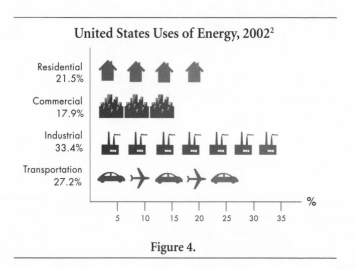

United States Uses of Energy, 2002[2]

Residential 21.5%

Commercial 17.9%

Industrial 33.4%

Transportation 27.2%

%

5   10   15   20   25   30   35

Figure 4.

Chapter 5 | Sources of Energy

Oil is the primary fuel source for the entire world.[1] It has huge advantages that make it useful in transportation, primarily its portability and its energy density—its ability to pack enough energy into a vehicle or airplane to provide satisfactory range.

However, along with the positives there are some negatives. Oil will soon become scarce. Over 60 percent of the world's remaining oil reserves are located in the Middle East, and our favored energy source is a polluting and greenhouse-gas-producing hydrocarbon.[2]

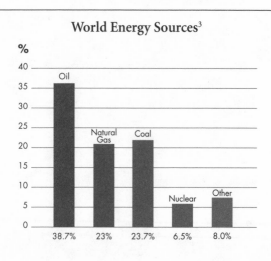

World Energy Sources[3]

Figure 5.

Natural gas is also a hydrocarbon, but is somewhat cleaner burning than oil. We don't expect natural gas to peak in world production as soon as oil; however, the use of natural gas is increasing. Many new central power plants are using it, thereby accelerating the time when its production will peak and it also will become a scarce resource.[4] Hydrogen can be made from natural gas for use in automobiles, but because natural gas is a finite resource with a production peak scheduled only a decade or two later than that of oil, this is too much of a short-term solution.[5]

Coal is available in abundance compared to the other two hydrocarbon fuels, but it is still a finite resource. Time estimates of remaining reserves range from two hundred years to three hundred years or more.[6] Coal is the dirtiest of the hydrocarbon fuels; it produces the most pollutants and greenhouse gases.[7]

Nuclear energy has not become as significant in power generation as scientists envisioned some thirty to fifty years ago, primarily due to concerns about safety and waste disposal issues.

Other sources of energy comprise only about 8 percent of our total energy usage. They include hydroelectric, biomass (vegetation, animal waste, etc.), wind, solar, and geothermal sources. Hydroelectric power is the primary component of this group and it has already achieved most of its potential because the good locations for power dams on rivers are already being utilized. Wind, solar, and other energy sources are in the infancy of their development, compared to oil, natural gas, coal, and nuclear sources.

Pros and Cons of Energy Sources

	Coal	Oil	Natural Gas	Nuclear	Hydro-electric	Biomass	Wind and Solar
Potential to be a major source of energy in future decades, exclusive of carbon dioxide (CO_2)	Good	Poor	Fair–Poor	Good	Poor	Unclear	Unclear
Suitability for direct use in transportation (portability)	Poor	Very Good	Fair–Poor	Poor	Poor	Poor	Poor
Ability to produce energy when needed	Good	Good	Good	Good	Good	Good	Poor
Geopolitical considerations	Good	Poor	Fair	Fair–Good	Good	Good	Good
Greenhouse emissions (CO_2)	Poor	Poor	Fair	Very Good	Very Good	Poor	Very Good
Present costs	Good	Good	Good	Good	Very Good	Poor	Fair–Poor

Figure 6.

Chapter 6

Transportation
Energy Requirements

Oil-based fuels provide over 97 percent of the fuel for transportation, because the transportation industries require certain key features in their energy sources.[1] Some of the more important considerations are portability, energy density, ease of handling, and safety.

Portability is critical. Automobiles, trucks, buses, ships, and aircraft cannot be connected or tethered to a source of energy. Imagine the implausible alternatives: cars with three-hundred-mile electrical extension cords and airplanes connected to natural gas hoses. There are only a few transportation exceptions that do not require portability, such as electric trains and subways that receive power through rails or electric lines running along the train's path. Therefore, either energy must be **stored** on-board or a source of energy must be **carried**.

**Transportation Requires
Energy Source Portability**

Figure 7.

Let's consider two ways that energy can be **stored** on vehicles: in batteries or as hydrogen.

Batteries provide current for an electric motor. After traveling some number of miles, these batteries must be recharged with energy. The source of this energy can be any or all of the sources we possess for generating electrical power, including coal, natural gas, oil, uranium, water, the sun, and the wind. Note that batteries can provide the energy portability that vehicles require while using energy sources other than oil-based fuels. However, there are still significant issues to resolve related to the use of battery-powered vehicles, including the problems of range, power, and weight.

Hydrogen should be thought of as a way to store energy, rather than as a source of energy, because it is not found in a usable form in nature the way oil, natural gas, and coal are. It must be extracted from a compound such as water or natural gas, and this manufacturing process is energy-intensive. Nevertheless, it *is* possible to separate hydrogen, store it on a vehicle, and convert it into the energy needed to drive the vehicle.

However, **carrying** a source of energy on a vehicle is the much more common way of supplying the energy than storing it. Early ships and trains generated steam by burning the wood they carried. Later these systems converted to coal and then to oil or products made from oil, including gasoline, diesel fuel, and bunker C ship fuel.

Energy density measures the amount of energy that can be packed into a certain space or weight. The usefulness of automobiles and airplanes, for example, would be compromised if they needed to carry large, heavy sources of energy. Try to imagine an airplane—or even a car—using wood or coal as a source of fuel. Energy density was a key reason for the shift from wood to coal, and then from coal to oil-based fuels; people felt that this quality made the new fuel "better" in each instance. Considerations of energy density include not only the fuel itself, but also the size and weight of the fuel container on the vehicle. Hydrogen, for example, must either be stored as a liquid below the extremely low temperature of -423°F or as a compressed gas at a pressure of a few thousand pounds per square inch.[2] A liquid hydrogen container must be extremely well insulated, and therefore, bulky. A compressed hydrogen container must be very strong, and consequently, heavy.

Ease of handling is another important consideration for energy used in transportation. It is much more convenient for vehicles to use a source of energy that is a fluid (this is defined as a liquid or gas) rather than a solid like coal or wood. Fluidity allows the fuel to be moved through small tubes on the vehicle to the point of combustion.

Electrical power can also be easily conveyed to the point of use through wires.

Safety must also be considered when evaluating fuels that might be alternatives to oil-based fuels. Gasoline itself is flammable and thus somewhat dangerous. However, possible alternative fuels like compressed natural gas or hydrogen have the potential to explode, both during refueling and in the inevitable traffic accidents.

It's easy to understand why oil-based fuels have become the mainstay of our transportation industries, after examining these four properties. It will be more of a challenge to replace oil than it would be to replace any other source of energy. Even after we decide on alternative sources, it will be a huge task to replace the Earth's 750 million vehicles and the fueling infrastructure.[3]

Section 2

Understanding Oil

Chapter 7

World Supply and
Demand for Oil

Small quantities of oil were produced and sold in the 1800s, but the Texas Spindletop gusher of 1901 marked the real beginning of the oil industry. Since then, the world's demand for oil has grown to a rate of over 82 million barrels per day.[1] We have used close to one-half of all the recoverable oil on our planet during this time.[2] To put this into perspective, this oil took at least 90 million years to create, and we have used about half of it during the lifetime of my ninety-five-year-old mother-in-law.[3] Because oil is being used for many more tasks than it was a hundred years ago, and because there are four times as many people on our planet as there were then, it isn't hard to understand why we are using up the remaining oil at a much faster rate than we did during the 1900s.[4]

To have a firm grasp of this situation, it is important to understand how oil is found and produced. Using sophisticated technologies, geologists identify likely spots for oil, then test wells are drilled, and oil fields are developed. Generally, it takes as much as a decade from the time a new oil site is discovered until a field is developed and producing significant quantities. The oil underground in porous rock formations moves under natural pressure toward the wells drilled into the formations. This oil generally flows quite freely for years or even decades, until the flow decreases as the field becomes depleted. Various techniques are used to coax additional oil from the field at this time, including fracturing the rock formations with pressure and acid or pumping water or carbon dioxide into nearby wells to force more oil to the producing wells. However, the production of the wells and the field as a whole always peaks and declines. When all the oil that can be economically recovered is gone, the field is abandoned.

Until fairly recently, the oil fields of the world have been capable of producing enough oil to satisfy our demand. There have been brief periods of interruption due to wars or imposed shipping limits, but the supply always existed to match the demand.

However, it has become increasingly difficult to find new reserves. World oil discoveries peaked more than fifty years ago and declined to around one-fourth of the peak by the 1990s. According to *The End of Oil*, the world has used 24 billion barrels of oil per year since 1995, while finding an average of only 9.6 billion barrels per year.[5] Today, the world uses three to four barrels of oil for every new barrel discovered.[6] Energy consultant corporation Wood Mackenzie reported that the "commercial value of oil and gas discovered over the past three years by the ten largest listed energy groups is running well below the amount they have spent on exploration."[7] We are coasting on existing fields without replacing our oil reserves.

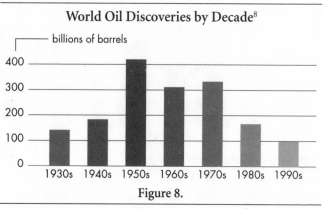

World Oil Discoveries by Decade[8]

billions of barrels

400
300
200
100
0

1930s   1940s   1950s   1960s   1970s   1980s   1990s

Figure 8.

The oil reserves-to-production (R/P) ratio is a measure that has been calculated for years for individual oil companies, oil fields, countries, and the entire world. The R/P ratio divides the proven oil reserves in the ground by current annual oil production. According to British Petroleum, the world R/P ratio for 2003 was 41 years.[9] That is, at the 2003 usage rate, the proven oil reserves would last forty-one years. This number is down from 43.7 years in 1989, because reserves have increased only 12.5 percent and production has increased by 20 percent since then.[10] The use of the R/P ratio has engendered a sense of complacency, prompting many to tell themselves, "We have over forty years of oil available, and we always seem to find more."

However, a more realistic school of thought states that we should focus on the time when world oil production peaks and begins its decline, rather than stressing the R/P ratio. L. B. Magoon, a geologist with the United States Geological Survey (USGS) who has mapped world oil fields for three decades, says that "are we running out of oil?" is the wrong question to pose. He says that instead we should ask, "When is the big rollover?"[11] The big rollover is defined as the time when world oil production peaks and rolls over into a permanent decline. After the peak, as production declines and demand continues to increase, oil will become scarcer each year. The year of worldwide peak oil production is found using the Hubbert's Peak calculation, which shows that the annual production from an oil field or a larger geographic area, when plotted as a graph, approximates a mountain or a bell-shaped curve, a curve common in statistics.

How is the curve calculated and drawn? First, researchers plot the known production by year. Although this may require some estimation, the figures are reasonably well known. Next, they estimate the total oil that can be recovered, which includes all the oil produced to date, as well as the projected amount that can be produced in the future. The total oil to be recovered, past and future, is represented by the area under the curve. Using the plot of past production and factoring in the total area beneath that curve, a computer can draw the rest of the curve, resulting in a bell-shaped plot that shows the peak. Then, we only have to note the year of the peak, after which production declines.[12] Approximately half of the total oil on the planet has been produced and consumed at the time the peak year occurs.

Hubbert applied his ideas to the lower forty-eight states of the United States in 1956 and predicted that oil production would peak between 1966 and 1972. The actual peak came in 1970, validating his approach.[13] The United States peak was 9.4 million barrels per day in 1970; production fell to 4.8 million barrels per day by 2002, a decline of 49 percent in thirty-two years, despite the addition of Alaskan production to our resources.[14]

The actual peak came in 1970, validating his approach.

Rise, Peak, and Decline of United States Oil Production[15]

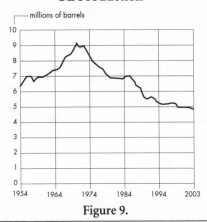

Figure 9.

Hubbert later applied his approach to the world and estimated the world's peak production would occur between 1990 and 2000. Other scientists have since used updated information to refine his calculations, estimating that the peak year for world oil production will occur as early as 2003 or as late as 2020.[16] Some of the estimates calculated are imminent, and none of the others is more than sixteen years in the future. Matt Simmons, oil industry expert and advisor, says, "Peaking of oil and gas will occur, if it has not already happened, and we will never know when the event has happened until we see it 'in our rearview mirrors.'"[17]

Although the Energy Information Administration (EIA) estimates that the oil production peak could come later, their researchers still state, "Our analysis shows that it will be closer to the middle of the twenty-first century than to its beginning. Given the long lead times required for significant mass-market penetration of new energy technologies, this result in no way justifies complacency about both supply-side and demand-side research and development."[18]

World Oil Production[19]

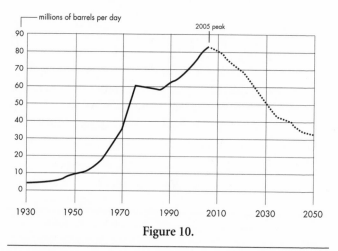

Figure 10.

Once we reach the year of peak production, world oil production will begin to decline. If we don't learn to use oil at a reduced rate or find alternatives, the demand for oil will continue to increase each year. The International Energy Agency (IEA) forecast for demand growth is 2.3 percent for 2005.[20] We can make assumptions about the rate of decline in production. Then, we can use these assumptions to create a graph that shows the increasing gap between demand and supply of oil for each year in the future, a measure of how scarce oil will be. Given the fairly optimistic assumption of a supply or production decline rate of 0.5 percent per year and a demand increase rate of only 1.5 percent per year, well under the current rate, just six years after the peak we will have an unfilled demand of over 10 million barrels per day. In eleven years, the shortfall will be over 25 percent of current demand. Recently, we have seen that the mere threat of being short 1 or 2 million barrels per day results in whopping price increases and world tensions. The predicted scarcity of oil will cause cars and planes to be permanently parked, unless we develop alternative fuels.

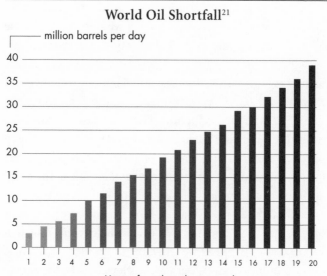

World Oil Shortfall[21]

million barrels per day

Years after oil production peak
Basis: supply declines at 0.5%/yr and demand grows at 1.5%/yr
Figure 11.

The important concept to grasp is that scarce oil and all the attendant problems will occur not when all the oil is gone, but when production can no longer keep up with demand, and this will be soon. Whether the peak and the beginning of declining production occur in 2005, 2007, or 2020, we must realize we have been procrastinating on developing robust fuel alternatives to oil. We should have begun years ago.

Chapter 8

Location of the World's Oil Supply

The oil industry really began in the United States, and the discovery and development of fields there fueled industry and an energy-intensive way of living during the twentieth century. The United States had sufficient oil for its operations during World War II, while Germany and Japan were beset with the difficulty of obtaining oil from beyond their borders. However, during this same period of time, the United States also developed habits that required large amounts of oil, with the result that it had to begin importing this fuel shortly after the Second World War to meet internal demand.[1] By 2002, the United States imported about 58 percent of its oil.[2] Today, the Middle East, containing over 63 percent of oil reserves, dominates the world oil supply.[3]

Today, the Middle East, containing over 63 percent of oil reserves, dominates the world oil supply.

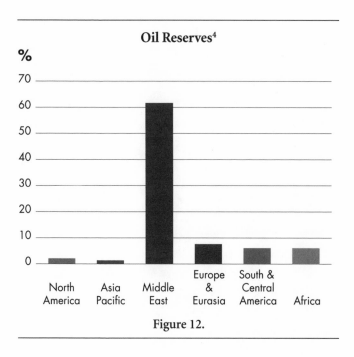

Figure 12.

Chapter 9

Oil Demand by Geographic Area

North America uses more oil than any other region of the world. In fact, in 2000, the United States, which contains only 5 percent of the world's population, used 25 percent of the oil produced in the world that year. Americans use two times the energy per capita of people living in Japan and Europe, and ten times that of the average world citizen.[1]

Developing nations, particularly China, are rapidly increasing their demand for oil. In 2004, China alone accounted for 40 percent of the rise in demand.[2]

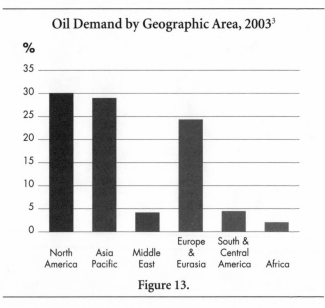

Oil Demand by Geographic Area, 2003[3]

Figure 13.

Section 3

Oil Alternatives

Chapter 10

Alternatives to Oil
for Transportation

We don't yet know which fuel alternatives have the potential to assume oil's role in transportation, but the facts all support a need to accelerate research for these alternative sources in order to reduce world dependence on hydrocarbon fuels. Citizens should have a basic understanding of the key issues surrounding the important fuel alternatives being considered.

Any viable alternative source of energy must exhibit a certain level of energy efficiency. If we use more energy in mining, growing, preparing, separating, collecting, or purifying a source than it will yield, then the source is of questionable value.

It is important to note that it is very difficult to predict the future cost of a product still in the research phase. If an alternative energy source meets all the other vital criteria, technology and industry often can reduce its costs well below any forecast we might make. Personal computers, automobiles, calculators, and television sets were all prohibitively expensive when first sold, but now are priced quite reasonably.

Any viable alternative source of energy must exhibit a certain level of energy efficiency.

Hydrogen-fueled vehicles have received a lot of press recently. This publicity contributes to a belief that an alternative to oil is at hand, leading to complacency in the search for other fuel sources. However, the problems hydrogen presents are daunting. Robert Service of the *Financial Times* wrote, "Recent reports from the United States National Academy of Sciences and the American Physical Society conclude that researchers face huge challenges in finding ways to produce and store hydrogen, convert it to electricity, supply it to consumers, and overcome safety concerns. The transition to a hydrogen economy, if it comes at all, will not happen soon."[1]

Hydrogen is not a source of energy found in nature in an almost-useable form like natural gas, oil, or coal. We should think of hydrogen as a storage system for energy, rather than as a source. It is an element that, when chemically combined with other elements, forms natural gas and water, among other substances. Energy-intensive processes are required to separate hydrogen from natural gas and especially from water. The major problem with obtaining hydrogen from natural gas is that natural gas production will peak not too many years after oil; it, too, is a limited resource.

The transition to a hydrogen economy, if it comes at all, will not happen soon.

Hydrogen can also be produced from water by electrolysis. In *The Hype About Hydrogen*, Joseph Romm notes that it takes about 1.4 units of energy to produce the hydrogen needed to generate 1.0 units of energy.[2] If we assume that the energy to perform the electrolysis would come from a central power plant using coal or natural gas fuel operating at 30 percent efficiency, overall it would take about 5 units of energy to produce 1 unit of hydrogen energy.[3] We must also remember that coal and natural gas are finite fossil fuels that will generate greenhouse gases when they are used to make the electrical energy to produce hydrogen.

Once produced, hydrogen can be liquefied by energy-intensive cooling to -423°F, at which point it becomes difficult to store and transport, because it requires a lot of insulation.[4] Hydrogen has a low energy density; even liquefied, it yields only about one-fourth the energy of gasoline from the same volume.[5] Hydrogen also can be used as a gas if compressed by energy-intensive multistage compressors to pressures of several thousand pounds per square inch. The requisite strong tanks to contain this gas are heavy. Hydrogen molecules are prone to leakage, quite easy to ignite, odorless, and very explosive; remember the Hindenburg dirigible.

Hydrogen has some advantages, however. It produces no greenhouse gases or pollution, is very efficient once aboard the vehicle, and can be produced from an abundant resource (water).

Assuming that the problems mentioned can be solved satisfactorily, hydrogen can be used by a vehicle's fuel cell to generate electrical power for an electric motor. Romm states, "Fuel cells convert the energy of a chemical reaction between a fuel such as hydrogen and an oxidant such as oxygen directly into electrical energy and heat."[6] A fuel cell operates generating no pollution.

Battery power is also an energy storage system, rather than a source of energy. A battery supplying energy to power a vehicle gives portability to all the sources of energy that can be used to generate electricity—a major advantage. These sources can be nuclear, hydroelectric, wind, and solar, all of which are nonpolluting.

> **A battery supplying energy to power a vehicle gives portability to all the sources of energy that can be used to generate electricity.**

However, the promise of battery power also brings with it the issues of range, power, and weight. These are critical, perhaps fatal, concerns for the use of batteries in aircraft. Batteries' weight and short range have also made them fairly unattractive for use in motor vehicles.

Natural gas can be compressed and used in vehicles in pressurized tanks. Of course, the compression process requires energy from some source. The weight of the tank and the shorter range of the vehicle are disadvantages compared with oil-based fuels. However, buses and other fixed-route vehicles operate now with compressed natural gas. Natural gas produces the greenhouse gas carbon dioxide (CO_2), but to a lesser degree than oil-based fuels. Perhaps the biggest issue with natural gas is that it also is a hydrocarbon destined to peak in world production not too many years after oil, even if we don't use large quantities of it for transportation.

Shale oil and oil sands are energy sources that have been known for many years. Production is expected to rise from

1 million barrels per day now to 3 million by 2015, meeting only a very small percentage of world oil demand, currently at 82 million barrels per day.[7] The principle drawback of these fuels is their low net energy payback; the energy required to process them is equal to a high percentage of the energy the finished product yields. Canada is now using natural gas—itself a finite and relatively clean energy source—to process oil sands. The processing method creates serious environmental and water use problems, and the combustion of the final product produces greenhouse gases.

Coal is not very practical in its solid form for use in ground vehicles, and is extremely impractical for air transportation. It can be liquefied, but the process requires a great deal of energy. The resulting fuel is still a hydrocarbon, and one of its products of combustion is CO_2, the most pervasive of the greenhouse gases. Coal is a finite nonrenewable source of energy, and accelerating its use will reduce the time before it, like oil, eventually becomes scarce.

Nuclear energy is a source that is difficult to imagine using directly in vehicles, given the current technology. It requires heavy shielding and presents radiation safety issues that make its implementation as a transportation fuel very problematic.

Solar and **wind** energy sources have potential to supply some of the electricity required for a battery-powered-vehicle industry. Direct solar energy depends on the sun shining on the vehicle when we want to go somewhere. This might be overcome to some extent with batteries, which still have their own weight and range problems. Solar cars do exist in experimental form, but the amount of solar energy that falls on a car yields very marginal performance.

Biomass, which includes trees and crops, is a fuel source with poor energy density in the natural state. Research also indicates that it would be energy inefficient. Corn has been used to make ethanol, a liquid fuel for vehicles, but Cornell University professor David Pimentel found that growing the corn and distilling the ethanol takes 1.7 times as much energy as the finished ethanol will produce.[8]

Hydroelectric power, energy created by a river, is impossible to use directly in vehicles.

Chapter 11

Energy Alternatives Needing
Research and Development

Focusing on research and development (R&D) in alternative fuel technologies is an extremely high priority for both the United States and the world. We must accelerate the rate at which we are solving the knotty technical issues that hinder some of the possible replacements for oil. Will hydrogen or electrically charged batteries or something else entirely power hundreds of millions of vehicles?

Governments must send the right signals to industry. Then government and industry must partner to produce solutions. Unfortunately, the complexities of the technical issues coupled with political pressures have made it difficult to formulate a plan to replace oil.

Successful corporations use a funnel-like approach to manage their product development efforts.[1] They have a somewhat loose selection process for choosing projects that receive initial funding. They know that they can flesh out the concept of a product and its prospects for success quite cheaply at an early phase, so they have many such projects in the idea and research phases. In later phases, these companies weed out or kill projects that appear unlikely to overcome their technical, cost, or marketing handicaps. Projects must meet predetermined criteria to advance to the next phase; thus the process generally allows for several pass/fail opportunities in the phases of a project's life. Since the costs of continuing a project increase exponentially as it moves from the research lab to prototypes and then on to full production, it becomes more important as time passes to be sure the final product will be successful before continuing to assign technical manpower and funds to it.

This funnel-like process permits a corporation to keep all its new product ideas, with their sales potential, technical risks, and costs, displayed constantly in full view. Thus, the organization can decide whether to continue to fund or to kill a project while comparing it to all the competing projects. Limited resources are allocated based on the best technical, cost, and marketing information available.

It is logical that we should apply this funnel model to organizing our direction and our funding efforts for research and development and the implementation of oil alternatives. Led by scientists and engineers, this approach would give focus to government and industry efforts, acting as a substitute for the technique of throwing money at whatever latest development has the biggest political splash potential. Technical experts would manage the pipeline of research and development efforts, providing sound recommendations for the allocation of resources. These experts would provide the government and industry with a constantly updated long-range road map of research efforts to ensure that money and technical manpower were devoted to oil alternatives with the best prospects for success.

We are already late in working aggressively to decide on alternatives to oil.

Let's look briefly at the difference between research and development. *Webster's* defines research as "careful, systematic, patient study and investigation into some field of knowledge, undertaken to discover or establish facts or principles."[2] Several times in this book, I have expressed the concern that we are already late in working aggressively to decide on alternatives to oil. This concern is due to the fact that many of the possible alternatives have technical issues so difficult that they require research, and research work by its very nature and definition is work whose completion date cannot be fixed. Research has already been conducted on some of the prospective alternatives for years without definitive answers. After problems of an alternative to oil are fairly well understood, that is to say, after research, development can begin with a much more predictable schedule.

Funnel Example of the Alternative Energy R&D Pipeline

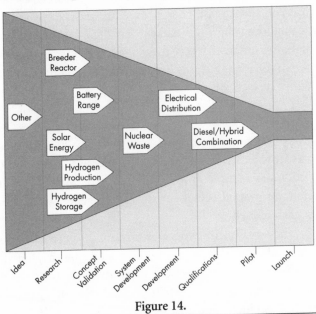

Figure 14.

Several energy alternatives currently require aggressive R&D:

Battery-Powered Vehicles

We must improve batteries' capability to provide satisfactory power and range in a package whose size and weight works well in vehicles. Successful R&D in the area of battery-powered vehicles could permit us to charge them at home or work and allow us to rely on all of the different sources of energy used to generate electricity, including the wind and the sun. The range problem might be mitigated by creating one standard-size universal battery, exchangeable in five minutes for a modest fee at any "service station."

Nuclear Energy

Battery power is a possible replacement for oil-based fuels in vehicles. If this revolution succeeded, we would need more central power stations, and nuclear power might be more acceptable than energy generated by hydrocarbon fuels.

The world now generates around 20 percent of its electrical power from fission nuclear power plants.[3] Relatively few nuclear power plants have been constructed worldwide in recent years, due to the fear of a nuclear accident and to the radioactive waste disposal problem. France generates 77 percent of its electrical power with nuclear energy, but the waste problem remains.[4] Breeder reactors, which both consume and produce fissionable fuel, have unsolved technical issues, but might mitigate the waste disposal problem by creating fuel during operation.[5]

Fission plants use uranium as a fuel. Like hydrocarbons, uranium is found in the ground and exists in finite amounts, but those total quantities worldwide are not as well known as oil reserves.

The other form of nuclear energy is fusion, the fusing of lightweight atoms. It has two big advantages: its fuel can be made from components of seawater and its waste decays to safe radioactive levels very quickly. However, nuclear fusion has some very challenging unsolved technical problems, such as confinement of the reaction by inertial or magnetic means.[6]

Hydrogen

As an alternative energy source, hydrogen has numerous technical issues, including transportation, storage, production onboard a vehicle, safety, distribution, and range. Scientists also have yet to discover how to efficiently produce hydrogen from a nonhydrocarbon source.

Coal

Coal is a hydrocarbon that is still fairly abundant. Some important issues related to coal are getting pollution control equipment incorporated in all coal-fired power plants, sequestering the CO_2 produced from its use, and making motor fuels from coal.

> Scientists have yet to discover
> how to efficiently produce
> hydrogen from a
> nonhydrocarbon source.

Electrical Energy Distribution

At least half of the electrical energy generated at power plants never reaches the point of use due to transmission losses.[7] Because of its incredible versatility, we aren't likely to stop using electrical power. However, we can reduce the waste. Possible areas for R&D work include generating power closer to the user, improving distribution systems, changing voltages, and using solar power for some of the load in homes.

Solar

Solar energy can be used to generate both heat and electricity. Current technology for solar-generated electricity results in costly power. It takes one to four years of operation just for a solar cell to generate the amount of energy required for its own production.[8] Only 0.07 percent of electrical power in the United States comes from solar cells.[9] There is a real need to lower the costs of the semiconductors used to turn the sun's rays into electrical power. The sun's energy is free, virtually infinite, and pollution free.

Wind

Wind energy equipment is being adopted at a steady rate. However, wind energy still represents a tiny fraction of the total electrical energy generated. Work on cost reductions and higher-volume manufacturing could accelerate its adoption. Like the sun, this energy source is free, vast, and clean.

Diesel Engines and Hybrids

Diesel engines provide substantially better fuel mileage than gasoline engines with very well-known technology. Hybrid vehicles combining an electric motor with a small gasoline engine are appearing in the automobile marketplace in increasing numbers recently and also demonstrate excellent fuel economy. Replacing gasoline engines with diesels in hybrid vehicles would combine the advantages of hybrid and diesel technologies, and would give even better fuel mileage. Diesel and hybrid vehicles still burn oil-based fuels, but they produce the same transportation with fewer barrels of oil. Both are bright spots on the development front, because they utilize available technologies that can prolong the longevity of oil.

A hybrid is under development that can be plugged in to provide sufficient charge to travel sixty miles.[10] Since the average American car is driven about thirty miles per day, a large percentage of vehicle travel can be undertaken using *no* oil-based fuels. A gasoline or diesel engine on-board provides instant switching to fuel to make a trip longer than sixty miles.

Methane Hydrates

Methane hydrates are estimated to contain twice as much carbon as all the other hydrocarbon sources of energy combined; they are a huge potential energy source.[11] Methane hydrates are methane gases trapped at a low temperature and high pressure, often on the ocean floor. Relatively little is known about how they might be obtained in large quantities for use as a fuel, so extensive R&D is required to determine whether they are a viable alternative. Methane is a major greenhouse gas when it escapes, and it produces CO_2 when it is burned.

Chapter 12 | Environmental Issues

The impending scarcity of oil alone is more than enough reason to research alternatives aggressively. However, many argue that environmental issues are just as critical.

Global warming is caused by an excess of greenhouse gases such as carbon dioxide (CO_2), and is exacerbated by the burning of hydrocarbon fuels. By definition, hydrocarbons include carbon in their makeup, and one of their products of combustion is CO_2. Natural gas produces the least CO_2, oil produces 1.4 times as much carbon dioxide as natural gas, and coal produces 1.8 times the amount.[1]

The greenhouse effect is caused by the accumulation of greenhouse gases in the Earth's upper atmosphere, inhibiting the radiation of heat from our planet. This causes a gradual warming of the earth with undesirable effects on the environment, raising the level of the oceans and changing the climate.[2] Although some questions about this effect still remain unanswered, we should not gamble our planet by doing nothing. If the accumulation of greenhouse gases were to become acute, it could take a century or more to reduce the excess amounts.[3]

Hydrocarbon fuels produce huge amounts of these gases as the result of combustion. Our use of coal, the dirtiest of the hydrocarbon fuels, produced 13 billion tons of CO_2 in 2001. The volume of this CO_2 is equivalent to the interior volume of 14 billion homes.[4] This quantity should be enough to make us think twice about energy solutions that require switching to greater use of coal simply because the world has at least a two-hundred-year supply available. Coal already plays a major role in electrical power generation in the United States, as well as in other parts of the world.

Energy Sources for Electrical Power Generation in the United States, 1998[5]

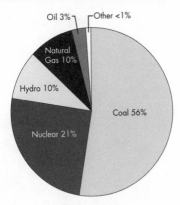

Figure 15.

Air pollution is another evil of hydrocarbon fuels. For example, it causes health problems and creates the need for expensive controls on vehicles and power plants. Lead, sulfur, sulfur dioxide, nitrogen oxide, and other pollutants are all emitted to varying degrees by the burning of these fossil fuels. Coal is by far the worst offender, especially in production of sulfur dioxide and particulates. Expensive controls are effective in reducing these emissions, and the more prosperous industrialized countries have made progress in recent decades. However, China and other developing countries are building coal-fired power plants at a rapid rate and manufacturing large numbers of automobiles. The primary interests of these countries lie in industrialization and increasing the number of consumer goods and exports, not in pollution control.[6]

Carbon Dioxide Emissions[7]

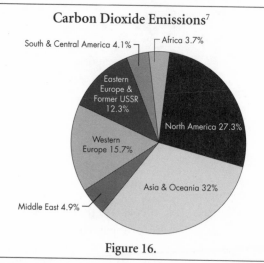

Figure 16.

Nuclear, wind, solar, and hydroelectric energy sources don't produce greenhouse gases or create air pollution problems. Of course, nuclear energy does have a serious radioactive waste disposal problem: its waste materials remain dangerously radioactive for over ten thousand years. In the United States alone, there are huge amounts of these materials that have not been put into permanent storage places. Why? Because people don't want these substances nearby, as it is perceived to be difficult to store them in such a way that they will not leak before they are no longer dangerous.

Section 4

Consequences and Steps for Action

Chapter 13

Life in a World without Alternatives to Oil

In any thorough examination of the oil shortage, it is important to discuss what will happen if we don't have viable alternatives fully developed when the oil supply can no longer meet demand.

Economic Consequences

- **Inflation** will be stimulated. Oil prices will increase dramatically to levels impossible to predict.

- **Standards of living** will deteriorate if oil becomes extremely scarce. Manufactured goods, air travel, automobile commuting, and vacations will eventually cost much more. Economies will suffer, because spending on oil-based products will reduce the amount of money left to spend on all other goods and services. We will have to change our habits drastically. As we begin to rely on other existing sources of energy, such as natural gas and coal, the increasing demand for them will raise their prices, making heating, air-conditioning, lighting, and manufacturing more expensive.

- The global **competitiveness** of countries that have been heavily dependent on cheap oil will be reduced. The American economy thrives on inexpensive fuel for transportation, distribution, and manufacturing. Removal of this crutch will make competing with low-cost-labor countries like China and India more difficult; those countries will be able to manufacture goods cheaply using low-pay workers, and the American economy won't be able to compensate using energy-intensive machines.

- The **balance of trade** will be adversely affected. The outflow of money from countries importing oil to those exporting it will be huge. For example, an increase of $60 per barrel over present prices for United States oil imports would mean an increase in cash payments to foreign countries of $249 billion per year, or an additional $846 per person. This would be added on to an already serious $600 billion current account balance.[1] Of course, much of this money would flow to the Middle East. In the *Financial Times*, Martin Wolf stated, "Unless trends change, ten years from now the United States will have fiscal debt and external liabilities that are both over 100 percent of GDP. It will have lost control over its economic fate."[2]

[The United States] will have lost control over its economic fate.

- Crash conversions to new and overdue technologies will lead to economic **inefficiencies.** Imagine the capital investment required to provide new service stations over just a few years for hydrogen or charged electric batteries. This is the sort of predicament that will occur if we wait too long to decide what alternative fuels and technology to phase in as oil's replacement.

  Consider another example: suppose gasoline cost so much that you bought an electric car while your SUV was still new, and no one wanted to buy your SUV. The economy *can* absorb such changes, as it always has, if they are begun early and phased in at a reasonable pace. Otherwise, funds for rapid and inefficient changes to new technologies will come out of our pockets.

- **Petrochemical** products depend on oil and natural gas. Up to this point, we have focused on the use of oil as a source of energy, especially for transportation. However, 7 percent of oil has a very important use as raw material for the petrochemical industry.[3] Have we thought about how we can get along without plastics, cleaning fluids, lubricating oil, adhesives, herbicides, cosmetics, pharmaceuticals, films, coatings, tires, packaging, synthetic fibers, insulation, inks, solvents, brake fluid, anti-icing agents, Velcro, and acids? Oil and natural gas are also required to produce nitrogen for the fertilizers that have made farming so incredibly productive.[4] Future generations will be angry with us for burning up the world's limited supply of oil in our cars and planes.

Political Consequences

- **Conflicts**—possibly even wars—will occur over which countries get to buy the remaining oil and what price should be charged. Even worse, there are serious moral issues with any country, including the United States, using military might to take oil from another country. The United States is perceived by many to wastefully have used more than its share of oil. It could find itself in serious competition with

Japan, China, India, and European nations—
buyers that might be favored over the United
States

- **Terrorists** will increasingly be able to disrupt
economies and peace. As oil becomes scarce,
terrorists might well take advantage of the
greater interruption they can cause by dam-
aging the delicate balance in oil and other
energy production and distribution systems.
They will discover more potential targets as
oil, natural gas, and coal have to be moved
greater distances to their markets.

Chapter 14 | Taking Action

Do we want to be at the mercy of another part of the world? Do we want our country to become less competitive? Do we want to become heavily indebted to other countries? The time has come to banish apathy and become more proactive in smoothing the transition from oil to energy alternatives.

Our efforts should involve both R&D and fuel conservation. Implementation in either category requires personal sacrifices, like paying more for fuels, agreeing to higher taxes, buying smaller cars, and spending personal income on energy-saving devices. These sacrifices may seem unattractive, but they are better than the alternative.

The time has come to banish apathy and become more proactive in smoothing the transition from oil to energy alternatives.

Research and Development

Here's what we shouldn't do: sit on our hands and hope that some huge new oil fields will be located. Most geology professionals feel that the world has been well searched, and that relying on big new discoveries is unwise and unrealistic. Oil discoveries have steadily decreased over the past fifty years, despite major improvements in the technologies used to find oil.

In the same vein, some people believe that drilling the Alaska National Wildlife Refuge (ANWR) would solve our oil problems. If ANWR is developed, it will not be functional for at least five years, and even then it will satisfy only 7 percent of the United States demand during its lifetime. ANWR encompasses about 1 percent of world oil reserves.[1] These amounts are not trivial, but they are also not a solution to scarce oil.

The United States has the technical and financial resources to complete huge, difficult projects successfully: consider the Marshall Plan and John F. Kennedy's ambition to send a man to the moon. America can play a leadership role in developing and implementing alternatives to oil; we just need the political will.

How to take action:

- Vote for policymakers who believe that government-industry partnerships can do the necessary research and development of alternative fuels.

- Insist on sound, long-range alternatives to oil, rather than settling on temporary fixes that lead to a false sense of security.

- Communicate via phone, letters, and e-mail with your senators and representatives to indicate that you want R&D accelerated, even if this initially negatively affects your income and comfort. One of many websites providing contact information for United States senators and representatives is http://www.visi.com/juan/congress.

- Stay informed on the status of alternatives to oil and discuss these options with your friends and neighbors.

America can play a leadership role in developing and implementing alternatives to oil.

Fuel Conservation

Why should we conserve fuels if they are eventually going to become scarce anyway? The answer is that conserving them will grant us more time to develop alternative fuels. Serious conservation efforts could slow or stop the growth rate of oil consumption and reduce the coming supply shortfall. However, this isn't happening yet. The Pogo cartoon read, "We have met the enemy and he is us." Nowhere is this statement truer than in our waste of fuels.

We can drive an SUV loaded with people and luggage twenty miles for the price of an ice cream cone.

How to take action:

- Drive less. The United States is the largest consumer of oil in the world, with over seventy-five cars for every one hundred men, women, and children. On a per-capita basis, we use twice as much oil as the people of Japan and Europe.[2]

  One reason for this is that oil is cheap at the retail level in the United States. We can drive an SUV loaded with people and luggage twenty miles for the price of an ice cream cone. For many years, gasoline prices in the United States have been about one-third of those in Europe. European countries tax gasoline heavily, which has resulted in the development and use of much more efficient vehicles. Europe's high gasoline prices have set into motion millions of individual decisions that have reduced fuel consumption. Wasteful use of oil-based products in the United States is an addiction, like alcoholism, which can be curbed. However, most American policymakers consider higher fuel taxes to be political death and are unlikely to support them unless the voters demand action.

We can also save fuel by working at home, living closer to work and school, riding public transportation, carpooling, walking, and riding bicycles. We can make a pledge to drive fewer miles each year.

- Drive more efficient vehicles. Diesel and hybrid vehicles present very real opportunities for conserving fuel. While they do use oil-based fuels, both use much less than most of the cars on the roads in America right now. Newly developed diesel technologies greatly reduce the objectionable attributes of diesel engines, such as noise, poor power, and black smoke.
- Fly less. Aviation in the United States consumes about 9 percent of the total transportation energy we use.[3] Air travel has become relatively

affordable. People travel to a distant vacation spot just for a long weekend or fly to another continent for a short business meeting. We just might have to cut back on such travel, and higher prices may be the most direct way to encourage frugality. We might also consider building more electric trains. In Europe and Japan they are well used, fast, and comfortable.

- Conserve other fuels. There are many ways we can conserve other fuel sources. We can insulate homes better, wear proper clothing, avoid excessive heating and cooling of homes and businesses, buy efficient lighting, and use local wind and solar energy.

- Once again, let your elected representatives know how you feel. We can vote for senators and representatives who are conservation-minded, and who are not beholden to industries that benefit from the status quo.

- Discuss conservation of fuels with your friends and neighbors. Set an example.

In 1933, Franklin D. Roosevelt spoke to the people, saying, "The only thing we have to fear is fear itself." But the statement didn't end there; he went on, criticizing "nameless, unreasoning, unjustified terror which paralyzes needed efforts to convert retreat into advance." These words inspired people to soldier on through the Great Depression. Fortunately, the oil crisis has not yet reached the magnitude of the depression, and we have yet to experience that terror. It is our duty and opportunity to prevent a severe oil crisis.

Notes

Chapter 1

1. David Goodstein, *Out of Gas* (New York: W. W. Norton & Co., 2004), 123.

2. Kenneth S. Deffeyes, *Hubbert's Peak: The Impending World Oil Shortage* (Princeton, NJ: Princeton University Press, 2001), 190.

Chapter 2

1. Energy Information Administration/International Energy Outlook 2004, "Reference Case Projections: World Energy Consumption, Gross Domestic Product, Carbon Dioxide Emissions, World Population," http://www.eia.doe.gov.

2. R. T. Kent, *Mechanical Engineers' Handbook,* ed J. K. Salisbury (New York: John Wiley & Sons, Inc., 1950), 3–50.

3. Ibid.

4. Ibid.

5. Richard Heinberg, *The Party's Over: Oil, War and the Fate of Industrial Societies* (Gabriola Island, BC: New Society Publishers, 2003), 142.

6. David Goodstein, *Out of Gas* (New York: W. W. Norton & Co., 2004), 96.

Chapter 3

1. *The Canadian Encyclopedia Online,* s.v. "The Origins of Petroleum," http://www.canadianencyclopedia.ca/index.cfm?PgNm=TCE&Params=J1SEC785989 (accessed September 23, 2004).

2. Richard Heinberg, *The Party's Over: Oil, War and the Fate of Industrial Societies* (Gabriola Island, BC: New Society Publishers, 2003), 50.

3. Paul Roberts, *The End of Oil: On the Edge of a Perilous New World* (Boston: Houghton Mifflin Company, 2004), 21.

4. David Goodstein, *Out of Gas* (New York: W. W. Norton & Co., 2004), 79.

5. Paul Roberts, *The End of Oil: On the Edge of a Perilous New World* (Boston: Houghton Mifflin Company, 2004), 271.

6. David Goodstein, *Out of Gas* (New York: W. W. Norton & Co., 2004), 28.

7. Energy Information Administration, "International Energy Outlook 2004," http://www.eai.doe.gov/oiaf/ieo/world.html.

8. "Unstoppable?" *The Economist,* August 21, 2004, 59.

9. Carola Hoyos, "Tough Choices for Oil Companies in the Quest to Head Off a Global Capacity Crunch," *Financial Times,* September 22, 2004.

10. Paul Roberts, *The End of Oil: On the Edge of a Perilous New World* (Boston: Houghton Mifflin Company, 2004), 146.

Chapter 4

1. Energy Information Administration, "Petroleum Flow 2002: Annual Energy Review 2002," http://www.eia.doe.gov/emeu/aer/diagram2.html.

2. Energy Information Administration, "Annual Energy Review 2002," http://www.eia.doe.gov/aer/diagram1.html.

Chapter 5

1. Energy Information Administration/International Energy Outlook 2004, "Reference Case Projections: World Energy Consumption, Gross Domestic Product, Carbon Dioxide Emissions, World Population," http://www.eia.doe.gov.

2. "BP Statistical Review of World Energy June 2004," British Petroleum, http://www.bp.com/subsection.do?categoryId=95&contentId=2006480 (accessed August 12, 2004).

3. Energy Information Administration/International Energy Outlook 2004, "Reference Case Projections: World Energy Consumption by Region and Fuel," http://www.eai.doe.gov

4. "Oil Crisis," Cooperative Research, http://www.cooperativeresearch.org/oil/oilcrisis.htm.

5. Bill Butler, "The Great Rollover Juggernaut," Durango Bill, http://durangobill.com/Rollover.html.

6. Kenneth S. Deffeyes, *Hubbert's Peak: The Impending World Oil Shortage* (Princeton, NJ: Princeton University Press, 2001), 173.

7. David Goodstein, *Out of Gas* (New York: W. W. Norton & Co., 2004), 33.

Chapter 6

1. Jean-Paul Rodrigue, "Transportation and Energy," Hofstra University, http://people.hofstra.edu/geotrans/eng/ch8en/conc8en/ch8c2en.html (accessed August 12, 2004).

2. Joseph J. Romm, *The Hype About Hydrogen* (Washington, DC: Island Press, 2004), 94.

3. Richard Heinberg, *The Party's Over: Oil, War and the Fate of Industrial Societies* (Gabriola Island, BC: New Society Publishers, 2003), 172.

Chapter 7

1. Jeffrey Ball, "As Prices Soar, Doomsayers Provoke Debate on Oil's Future," *Wall Street Journal,* September 21, 2004.

2. David Goodstein, *Out of Gas* (New York: W. W. Norton & Co., 2004), 28.

3. Kenneth S. Deffeyes, *Hubbert's Peak: The Impending World Oil Shortage* (Princeton, NJ: Princeton University Press, 2001), 16.

4. World population in 1900 was about 1.6 billion and is now about 6.4 billion. Joseph L. Zachary, "World Population," Krell Institute, http://krellinst.org/UCES/Demo/ScientificComputing/uces-1/uces-1/body-uces-1.html (accessed September 27, 2004).

5. Paul Roberts, *The End of Oil: On the Edge of a Perilous New World* (Boston: Houghton Mifflin Company, 2004), 51.

6. Richard Heinberg, *The Party's Over: Oil, War and the Fate of Industrial Societies* (Gabriola Island, BC: New Society Publishers, 2003), 108.

7. James Boxell, "Oil groups failing to meet costs of new finds," *Financial Times,* October 11, 2004.

8. C. J. Campell, "World Oil Discoveries," in *The Party's Over: Oil, War and the Fate of Industrial Societies* (Gabriola Island, BC: New Society Publishers, 2003), 108.

9. "BP Statistical Review of World Energy June 2004," British Petroleum, http://www.bp.com/subsection.do?categoryId=95&contentId=2006480 (accessed August 12, 2004).

10. Ibid.

11. L. B. Magoon, "Are We Running Out of Oil?" United States Geological Survey, http://geopubs.wr.usgs.gov/open-file/of00-320/.

12. Kenneth S. Deffeyes, *Hubbert's Peak: The Impending World Oil Shortage* (Princeton, NJ: Princeton University Press, 2001), 133.

13. Ibid, 4.

14. Energy Information Administration, "Crude Oil Production and Crude Oil Well Productivity, 1954–2003," Table 5.2, http://www.eia.doe.gov.

15. Ibid.

16. L. B. Magoon, "Are We Running Out of Oil?" United States Geological Survey, http://geopubs.wr.usgs.gov/open-file/of00-320/.

17. Paul Roberts, *The End of Oil: On the Edge of a Perilous New World* (Boston: Houghton Mifflin Company, 2004), 60.

18. John H. Wood, Gary R. Long, and David F. Morehouse, "Long-Term World Oil Supply Scenarios: The Future Is neither as Bleak or Rosy as Some Assert," (Energy Information Administration, 2004), http://www.eia.doe.gov.

19. C. J. Campbell, "Forecasts of Future Oil Output, 2004," Hubbert Peak of Oil Production, http://www.hubbertpeak.com/curves.htm (accessed October 7, 2004).

20. Kevin Morrison, "Crude oil prices recover most of their early falls," *Financial Times,* July 4, 2004.

21. Annual shortfall calculated assuming demand and supply both at 82 million bbl/day in year zero, then supply declining at 0.5%/year and demand increasing at 1.5%/year thereafter.

Chapter 8

1. Paul Roberts, *The End of Oil: On the Edge of a Perilous New World* (Boston: Houghton Mifflin Company, 2004), 41.

2. Calculated from the flow sheet by adding three categories of imports and dividing that sum by petroleum consumption. Energy Information Administration, "Petroleum Flow 2002: Annual Energy Review 2002," http://www.eia.doe.gov/emeu/aer/diagram2.html.

3. "BP Statistical Review of World Energy June 2004," British Petroleum, http://www.bp.com/subsection.do?categoryId=95&contentId=2006480 (accessed August 12, 2004).

4. Ibid.

Chapter 9

1. Paul Roberts, *The End of Oil: On the Edge of a Perilous New World* (Boston: Houghton Mifflin Company, 2004), 15.

Bill Butler, "The Great Rollover Juggernaut," Durango Bill, http://durangobill.com/Rollover.html.

2. Peter Coy, "The Trouble with Gushing Oil Demand," *BusinessWeek,* April 26, 2004.

3. "BP Statistical Review of World Energy June 2004," British Petroleum, http://www.bp.com/subsection.do?categoryId=95&contentId=2006480 (accessed August 12, 2004).

Chapter 10

1. Robert F. Service, "A dream of a hydrogen economy," *Financial Times,* August 13, 2004.

2. Joseph J. Romm, *The Hype About Hydrogen* (Washington, DC: Island Press, 2004), 75.

3. Ibid.

4. Ibid., 94.

5. Robert F. Service, "A dream of a hydrogen economy," *Financial Times,* August 13, 2004.

6. Joseph J. Romm, *The Hype About Hydrogen* (Washington, DC: Island Press, 2004), 23.

7. James Cox, "Canada Drips with Oil, but It's Tough to Get At." *USA Today,* September 7, 2004, http://yahoo.com/news?tmpl=stor y&cid=677&u=/usatoday/20040907/bs_usatoday.

8. 131,000 btu needed to make a gallon of ethanol with an energy value of 77,000 btu. 131,000/77,000 = 1.7. Richard Heinberg, *The Party's Over: Oil, War and the Fate of Industrial Societies* (Gabriola Island, BC: New Society Publishers, 2003), 156.

Chapter 11

1. Michael E. McGrath, Michael T. Anthony, and Amram R. Shapiro, *Product Development: Success through Product and Cycle-Time Excellence* (Boston: Butterworth-Heinemann, 1992), 188, 195.

Robert G. Cooper, Scott J. Edgett, and Elko J. Kleinschmidt, *Portfolio Management for New Products* (Cambridge, MA: Perseus Publishing, 2001), 12, 190.

2. *Webster's New World Dictionary,* 2nd ed., s.v. "Research."

3. Richard L. Garwin, "Can the World Do without Nuclear Power? Can the World Live with Nuclear Power?" Federation of American Scientists, http://www.fas.org/rlg/010409-nci.htm (accessed April 9, 2001).

4. Richard Heinberg, *The Party's Over: Oil, War and the Fate of Industrial Societies* (Gabriola Island, BC: New Society Publishers, 2003), 133.

5. "What is a breeder reactor?" EWS, http://www.cen.uiuc.edu/~comberia/introduction.html (accessed September 4, 2004).

6. "Nuclear fusion," Radwaste.org, http://www.radwaste.org/fusion.htm (accessed September 4, 2004).

7. Paul Roberts, *The End of Oil: On the Edge of a Perilous New World* (Boston: Houghton Mifflin Company, 2004), 226.

8. Vanessa Houlder, "Slow Dawn for the Rising Sun," *Financial Times,* June 25, 2004.

9. Otis Port, "Another Dawn for Solar Power," *BusinessWeek,* September 6, 2004, 94.

10. "Why the Future is Hybrid," *The Economist,* December 4, 2004, 26.

11. "Methane in the Deeps," American Litoral Society, http://www.sealitsoc.org/newsletter/methane.htm (accessed September 13, 2004).

Chapter 12

1. "Natural Gas and the Environment," NaturalGas.org, http://www.naturalgas.org/environment/naturalgas.asp#emission (accessed October 12, 2004).

2. Vijay V. Vaitheesvaran, *Power to the People,* (New York: Farrar, Straus and Giroux, 2003),127.

3. Ibid., 153.

4. Calculations made using 2001 world coal consumption of 95.9 quadrillion btu, converting to tons of coal, using 2.86 tons CO_2/ton of coal to calculate 13 billion tons of CO_2 in one year, then using 0.116 pounds of CO_2 per cubic foot and 2,000 cubic feet per house to calculate the number of houses. Energy Information Administration/International Energy Outlook 2004, "Reference Case Projections: World Energy Consumption, Gross Domestic Product, Carbon Dioxide Emissions, World Population," http://www.eia.doe. gov.

"Appendix E: Common Conversion Factors," Pacific Rim Energy and Environmental Network, http://www.preen.org/eiagg97/ appe.html (accessed November 1, 2004).

5. Richard Heinberg, *The Party's Over: Oil, War and the Fate of Industrial Societies* (Gabriola Island, BC: New Society Publishers, 2003), 61.

6. Paul Roberts, *The End of Oil: On the Edge of a Perilous New World* (Boston: Houghton Mifflin Company, 2004), 122.

7. "Carbon Emissions, 2002," Energy Information Administration, http://www.eia.doe.gov/pub/international/ielaf/tableh1.xls.

Chapter 13

1. Martin Wolf, "America Is Now on the Comfortable Path to Ruin," *Financial Times,* August 18, 2004.

2. Ibid.

3. Kenneth S. Deffeyes, *Hubbert's Peak: The Impending World Oil Shortage* (Princeton, NJ: Princeton University Press, 2001), 164.

4. Ibid., 165.

Chapter 14

1. Per reference, ANWR could produce 1.4 million bbl/day, and this figure divided by the current United States usage of 20 million bbl/day gives 7 percent. The ANWR expected reserve of 10.4 billion bbl is near 1 percent of the estimated world reserves, which number in the 1.0 trillion bbl range. United States Department of Interior, "ANWR Oil Reserves Greater Than Any State," htpp://doi.gov/news/030312.htm.

2. Jean-Paul Rodrigue, "Transportation and Energy," Hofstra University, http://people.hofstra.edu/geotrans/eng/ch8en/conc8en/ch8c2en.html (accessed April 13, 2004).

3. Energy Information Administration, "Users of Energy," http://www.eia.doe.gov/kids/consumption/transportation.html.

Bibliography

Akst, Daniel. "What If a Candidate Told the Truth about Oil?" *New York Times,* August 24, 2004.

American Coalition for Ethanol. "What is Fuel Ethanol?" American Coalition for Ethanol. http://www.ethanol.org/whatisethanol. html.

American Litoral Society. "Methane in the Deeps." American Litoral Society. http://www.sealitsoc.org/newsletter/methane.htm (accessed September 13, 2004).

Ball, Jeffrey. "As Prices Soar, Doomsayers Provoke Debate on Oil's Future." *Wall Street Journal,* September 21, 2004, Vol. CCXLIV No. 57.

Becker, Gary S. "Let's Make Gasoline Prices Even Higher." *Business-Week,* May 31, 2004.

Boxell, James. "Oil Groups Failing to Meet Costs of New Finds." *Financial Times,* October 11, 2004.

British Petroleum. "BP Statistical Review of World Energy June 2004." *British Petroleum.* http://www.bp.com/subsection.do? categoryId=95&contentId=2006480.

Butler, Bill. "The Great Rollover Juggernaut." Durango Bill. http:// durangobill.com/Rollover.html.

Campbell, C. J. *The Coming Oil Crisis.* Milhac, France: Multi-Science Publishing Company & Petroconsultants S.A., 1997.

———. "Forecasts of Future Oil Output." HubbertPeak.com. http://www.hubbertpeak.com/curves.htm.

Campbell, Colin J. and Jean H. Laherrere. "The End of Cheap Oil." *Scientific American,* March 1998.

Canadian Encyclopedia. "The Origins of Petroleum." http://www.canadianencyclopedia.ca/index.cfm?PgNm=TCE&Params=J1 SEC785989.

Carey, John. "Global Warming." *BusinessWeek,* August 16, 2004.

Cooper, Robert G., Scott J. Edgett, and Elko J. Kleinschmidt. *Portfolio Management for New Products.* Cambridge, MA: Perseus Publishing, 2001.

Cooperative Research. "Oil Crisis." Cooperative Research. http://www.cooperativeresearch.org/oil/oilcrisis.htm.

Cox, James. "Canada Drips with Oil, but It's Tough to Get At." *USA Today,* September 7, 2004, http://yahoo.com/news?tmpl=story &cid=677&u=/usatoday/20040907/bs_usatoday.

Coy, Peter. "The Trouble with Gushing Oil Demand." *BusinessWeek,* April 26, 2004.

"Crude Oil Production and Crude Oil Well Productivity, 1954–2003." Table 5.2. http://www.eia.doe.gov.

Deffeyes, Kenneth S. *Hubbert's Peak: The Impending World Oil Shortage.* Princeton, NJ: Princeton University Press, 2001.

Dillon, William. "Gas (Methane) Hydrates—A New Frontier." United States Geological Survey. http://marine.usgs.gov/fact-sheets/gas-hydrates/title.html.

Energy Information Administration. "Appendix E: Common Conversion Factors." http://www.preen.org/eiagg97/appe.html.

———. "Energy Consumption by Sector Overview." Figure 2.1a. http://www.eia.doe.gov.

———. "Energy in the United States: 1635–2000." http://www.eia.doe.gov/emeu/aer/eh/intro.html.

———. "International Energy Outlook 2004." http://www.eia.doe.gov/oiaf/ieo/world.html.

———. "Milestones." http://www.eia.doe.gov/kids/milestones/index.html.

———. "Petroleum Flow 2002: Annual Energy Review 2002." http://www.eia.doe.gov/emeu/aer/diagram2.html.

———. "Users of Energy." http://www.eia.doe.gov/kids/consumption/transportation.html (accessed August 13, 2004).

———. "What is Energy?" http://www .eia.doe.gov/kids/whatsenergy.html.

———. "World Energy and Economic Outlook." http://www.eia.doe.gov.

Energy Information Administration/International Energy Outlook 2004. "Reference Case Projections: World Energy Consumption, Gross Domestic Product, Carbon Dioxide Emissions, World Population." http://www.eia.doe.gov (accessed August 11, 2004).

EWS. "What is a breeder reactor?" EWS. http://www.cen.uiuc.edu/ ~comberia/introduction.html.

"Forecasts of Future Oil Output." HubbertPeak.com. http://www. hubbertpeak.com/curves.htm.

Foss, Brad. "Oil Prices Increase as Supply Remains Tight." *Austin American-Statesman,* September 23, 2004.

"Four Sensible Ways to Break Our Dependence on Foreign Oil." *BusinessWeek,* August 30, 2004.

Garwin, Richard L. "Can the World Do without Nuclear Power? Can the World Live with Nuclear Power?" Federation of American Scientists. http://www.fas.org/rlg/010409-nci.htm.

Geewax, Marilyn. "Are the days of cheap oil running out?" *Austin American-Statesman,* September 19, 2004, sec. J.

George, Nicholas and James Mackintosh. "Sweden gears up for a tax drive against Volvo's SUVs." *Financial Times,* September 1, 2004.

Goodstein, David. *Out of Gas.* New York: W. W. Norton & Co., 2004.

Heinberg, Richard. *The Party's Over: Oil, War and the Fate of Industrial Societies.* Gabriola Island, BC: New Society Publishers, 2003.

Hight, Bruce. "What Will We Do When the Oil Wells Run Dry?" *Austin American-Statesman,* June 27, 2004, sec. H.

Hoffman, Peter. *Tomorrow's Energy: Hydrogen, Fuel Cells, and the Prospects for a Cleaner Planet.* London: MIT Press, 2001.

Houlder, Vanessa. "Slow Dawn for the Rising Sun." *Financial Times,* June 25, 2004.

Hoyos, Carola. "Tough Choices for Oil Companies in the Quest to Head Off a Global Capacity Crunch." *Financial Times,* September 22, 2004.

International Energy Agency. *World Energy Outlook 2001.* International Energy Agency, 2001.

Ivanhoe, L. F. "King Hubbert—Updated." M. King Hubbert Center for Petroleum Supply Studies. http://hubbert.mines.edu/news/Ivanhoe_97-1.pdf.

Kent, R. T. *Mechanical Engineers' Handbook.* Edited by J. K. Salisbury. New York: John Wiley & Sons, Inc., 1950.

Klare, Michael T. *Resource Wars.* New York: Henry Holt and Company, 2001.

Leeb, Stephen. *The Oil Factor.* New York: Warner Business Books, 2004.

Loo, Felicia. "China's Oil Thirst Changes Global Flows." Reuters, May 20, 2004. http://www.forbes.com/business/energy/newswire/2004/05/12/rtr1368346.html.

Magoon, L. B. "Are We Running Out of Oil?" United States Geological Survey. http://geopubs.wr.usgs.gov/open-file/of00-320/.

McGrath, Michael E., Michael T. Anthony, and Amram R. Shapiro. *Product Development: Success through Product and Cycle-Time Excellence.* Boston: Butterworth-Heinemann, 1992.

Morrison, K., C. Hoyos, and Roula Khalaf. "Terror Attacks, Capacity Shortages and a Herd of Speculators: How Can OPEC Bring Calm to the World Oil Market?" *Financial Times,* June 3, 2004.

Morrison, Kevin. "Crude Oil Prices Recover Most of Their Early Falls." *Financial Times,* July 4, 2004.

Murphy, Cait. "Why $3-a-Gallon Gas is Good for America." *Fortune,* June 28, 2004.

Nakicenovic, Nebojsa, Arnulf Grubler, and Alan McDonald. *Global Energy Perspectives.* Cambridge, UK: Cambridge University Press, 1998.

NaturalGas.org. "Natural Gas and the Environment." NaturalGas. org. http://www.naturalgas.org/environment/naturalgas. asp#emission.

Port, Otis. "Another Dawn for Solar Power." *BusinessWeek,* September 6, 2004.

Radwaste.org. "Nuclear fusion." Radwaste.org. http://www.rad-waste.org/fusion.htm.

Reed, Stanley. "Cheap Oil? Forget About It." *BusinessWeek,* March 8, 2004.

Reitzle, Wolfgang. "We Need a New Fuel to Power the World Economy." *Financial Times,* August 11, 2004.

Roberts, Paul. *The End of Oil: On the Edge of a Perilous New World.* Boston: Houghton Mifflin Company, 2004.

Rodrigue, Jean-Paul. "Transportation and Energy." Hofstra University. http://people.hofstra.edu/geotrans/eng/ch8en/conc8en/ch8c2en.html.

Romm, Joseph J. *The Hype About Hydrogen.* Washington, DC: Island Press, 2004.

Service, Robert F. "A Dream of a Hydrogen Economy." *Financial Times,* August 13, 2004.

Simmons, Matthew R. "Energy Prices and Energy Fundamentals: Is There a Link?" Simmons and Co. International. http://www.simmonsco-intl.com.

Third World Traveler. "Oil Watch." Third World Traveler. http://www.thirdworldtraveler.com/Oil_watch.html.

———. "World Oil Facts." Third World Traveler. http://www.thirdworldtraveler.com/Oil_watch/World_Oil%20_Table.html.

United States Department of Interior. "ANWR Oil Reserves Greater Than Any State." htpp://doi.gov/news/030312.htm.

"Unstoppable?" *The Economist,* August 21, 2004.

Vaitheesvaran, Vijay V. *Power to the People.* New York: Farrar, Straus and Giroux, 2003.

Varchaver, Nicholas. "How to Kick the Oil Habit." *Fortune,* August 23, 2004.

Welch, David, Cathleen Kerwin, John Carey, and Ronald Grover. "California Rules, Detroit Quakes." *BusinessWeek,* July 12, 2004.

"Why the Future is Hybrid." *The Economist,* December 4, 2004.

Williams, James L., and A. F. Alhajji. "The Coming Energy Crisis?" Energy Economics Newsletter. http://www.wtrg.com/Energy Crisis/index.html (accessed May 20, 2004).

Williscroft, Robert G. "Nuclear Waste and Breeder Reactors—Myth and Promise." Nuclear Waste and Breeder Reactors—Myth and Promise. http://www.argee.net/DefenseWatch/Nuclear% 20Waste%20and%20Breeder%20Reactors.htm.

Wolf, Martin. "America is Now on the Comfortable Path to Ruin." *Financial Times,* August 18, 2004.

Wood, John H., Gary R. Long, and David F. Morehouse. "Long-Term World Oil Supply Scenarios: The Future Is Neither as Bleak or Rosy as Some Assert." Energy Information Administration. http://www.eia.doe.gov.

Zachary, Joseph L. "World Population." Krell Institute. http://www. krellinst.org/UCES/Demo/ScientificComputing/uces-1/uces-1/body-uces-1.html (accessed September 27, 2004).

Index

* Italicized page references indicate figures